Ghost Words

A 10-Minute Daily Habit to Catch AI Fabrications in Your Working Documents

BANDAR ALSHARFI

Ghost Words: A 10-Minute Daily Habit to Catch AI Fabrications in Your Working Documents

Published by CodeDebate Press, an imprint of RB-Ventures LLC.

Every case described in this book is a composite, assembled from patterns observed across many organizations. None is an account of a single real organization, and no organization is the subject of any example in these pages. Any resemblance to the circumstances of a specific organization reflects how common these patterns are.

ISBN 979-8-99733538-0 (Kindle)
ISBN 979-8-99733539-7 (paperback)

Quantity sales. Special discounts are available on bulk purchases by corporations, associations, and educational institutions. For details, contact the publisher at codedebate.com.

First edition, 2026

CodeDebate Press · codedebate.com

Contents

PART III. SPOTTING THE SPECIFIC HALLUCINATION

These chapters detail common fabrications like numbers and citations that require specific checks.

PART IV. MANAGING THE ERROR ONCE FOUND

These chapters cover correction, uncertainty, and structural fixes once an error is found.

PART V. SYSTEMATIZING ACCURACY ACROSS TEAMS

These chapters move from individual checks to household and team logs for systemic accuracy.

ALSO IN THIS BOOK

Before You Start: Ten Minutes and Why It Matters

The people and companies in this book are invented for illustration, and the first-person episodes are composites — real patterns from my working life, compressed and stripped of anything a reader could check.

Before You Start: Ten Minutes and Why It Matters

Marcus had been staring at the vendor comparison for three hours. The spreadsheet was immaculate. Every row matched and every total summed. The recommendation at the bottom glowed with the quiet certainty of a job finally done. It was 4:47.

He attached it to an email, typed "Recommended vendor selected" in the subject line, and hit send. Then he stood up, put on his coat, and walked out of the office. By 5:15 he was on the train. Tuesday morning he was in a conference room watching Dana the COO scroll through his file in silence.

She stopped on row fourteen. "This pricing is for the new platform only." Marcus felt his throat tighten. The legacy module his team still used wasn't included in the quote. The vendor's rep with the flawless demo had buried that detail in a footnote Marcus hadn't opened. Dana closed the laptop. "We need to rerun this."

Marcus did not skip the work. He skipped the last ten minutes. He'd spent three weeks gathering data, then sent it without switching his head from hunting mode to judging mode.

In my experience, the gathering feels like progress and the pause feels like delay. It's not. The ten minutes is where you ask what you're missing, not what you've found.

Priya works two desks over. She had a similar file due the same week, and she'd done the same three weeks of research. At 4:40 on Friday she set a timer for ten minutes. She stood up and read her own summary aloud to an empty chair. The gap jumped out immediately. She'd been comparing vendors for the new platform, assuming the legacy module was included because the sales rep said it was "compatible." It was not bundled. When the timer rang at 4:50, she sat back down, fixed line fourteen, and sent the file two minutes later. Dana approved it Monday morning without a meeting.

The timer is not a productivity trick. It's a perspective shift. After three weeks inside the details, you step back and ask what would make the recommendation fail in the first month.

I learned to do this after a report of my own went sideways. I'd checked every number twice. I hadn't checked whether the numbers answered the question anyone had asked. The ten minutes is for that second check. It's not for proofreading. It's for making sure you're solving the right problem.

Here is how to run it. Set a timer for ten minutes and step away from the screen. Close the research tabs. Look at your work as if you're the person who has to approve it, not the person who did it. State your actual recommendation in one sentence and name the assumption that could sink it. Note one fact you still do not know about the project.

If you cannot write the recommendation in one sentence, you're not ready to send. If you cannot name the fatal assumption, you haven't looked hard enough. If you cannot find one gap, look for a second.

Marcus now does the ten minutes before every major send. He says it feels like leaving money on the table when he catches nothing, and that's the point. The ten minutes is a better bet than trusting tired eyes after a long week.

Pick the biggest open item on your desk. Set a timer for ten minutes. Close your notes and write one sentence stating what you'll actually decide. Name the assumption that would make that decision fail. Put the paper in your drawer and read it after you finish the work. You haven't started the task. You've done the ten minutes.

PART I

Diagnosing the Empty Claim

The Fact That Wasn't in the Story

Marcus slid the brochure across the conference table and spun it so the numbers faced you. It was the StackFlow case study he'd emailed about that morning. Apex Distribution had cut picking errors by sixty percent in ninety days, and the before-and-after bar graph took up half the page. Marcus tapped the short red bar and said, "They did it with the dashboard. No new hardware, no extra headcount."

You could see he'd already decided.

Dana, who ran operations for Brightline, picked up the sheet and read the quote from Apex's VP twice before she turned it over. "What month did they go live?" Marcus checked his notes. "March." Dana nodded. "And when did Apex close the Denver warehouse?" You remembered the announcement from the same logistics brief you both read every Tuesday. Apex had consolidated west coast inventory into Portland that March. They'd moved half their SKUs to a third-party handler and cut the product line that needed the most manual work. Picking errors dropped because there was less complex picking to do, and the dashboard only counted what remained.

Marcus looked at the brochure again, but the warehouse closing wasn't in there.

Every case study is a story, and stories have authors who pick where the narrative starts and what counts as a cause. StackFlow started the clock the week their software went live, and they chose the metric that looked best. They didn't lie; they left out the warehouse closing, the SKU reduction, and the product-line trim because those details made the software look smaller. The missing fact wasn't hidden. It was assigned to a different story in a different department.

You see this everywhere once you start looking. The regional manager who doubled output did it in a month when two competitors shut down for maintenance. The new CRM didn't speed up the sales cycle; the company stopped serving small accounts and only called enterprise prospects. The training program that reduced turnover coincided with a raise that hit the same quarter. None of these facts are secret, but they're kept out of the official story because they complicate the moral. I have sat in rooms like that one — where the real cause was a layoff, and the team pinned the result on a new logo anyway.

The moral is usually the same: buy the product, adopt the method, copy the leader. The missing fact is the one that keeps your checkbook closed. I have signed contracts on stories this thin, and I have never been glad I did.

So how do you find it? Check the edges, because stories have boundaries and facts leak across them. Ask what happened just outside the frame, and ask what changed in the same month that had nothing to do with the hero. Ask what the client stopped doing. If a vendor claims their scheduling tool doubled output, find out whether the client also added a second shift, changed the product mix, or fired a slow supervisor. The fact that breaks the story is usually sitting in an adjacent paragraph of an annual report, a trade brief, or a hallway conversation with someone who left the company.

The other place to look is inside the metric itself. Sixty percent fewer errors sounds precise, but errors of what? If Apex stopped carrying fragile glassware and moved to sealed pallets, the error rate for breakage would vanish even if workers were still guessing. The dashboard might've done nothing but color-code the guesswork.

You don't need to prove malice. Most storytellers believe their own ending. The StackFlow rep who printed that brochure probably thinks the software caused the drop because nobody told him about the consolidation. He wasn't in the room when Apex's operations team made the real decision. He inherited a happy client and a clean graph, and he filled in the gap with the only cause he had permission to sell. That means finding the missing fact is your job.

Find a case study or vendor pitch that impressed you in the last six months. Write down the result they claimed and the month it happened. Now find one other major change that company made in the same quarter. A restructure, a layoff, a product-line cut, a new contractor. Ask yourself whether that change could explain the result without the vendor's help. If it could, you've found the fact they left out. File it next to the brochure before you spend anything.

Why the Machine Fills the Gap

Marcus in procurement had fifteen minutes before the review. He pasted the supplier's name and a three-sentence brief into the prompt. The model produced twelve pages of risk assessment in thirty seconds. He'd added his cover line and hit send. That afternoon Dana read the first page, skimmed the third, and forwarded it to legal. Her note said one word: "Thorough."

Nobody asked Marcus if the supplier was safe. They asked for a document that looked complete. The machine didn't create that problem. It simply removed the friction that once slowed empty work.

Before these tools, Marcus would have opened last year's file, changed the names and the dates, and sent the same skeleton back out the door. He'd still have submitted a document nobody planned to read. The difference is speed. Now a hollow request produces polished hollow prose in seconds. I've done the same thing. Early in my career I wrote weekly reports that went straight to a folder, bloated and useless, exactly what the system wanted. The folder was the customer. The machine is simply a faster writer for the same folder.

The gap is not a technology problem. It is an incentives problem. Dana asks for twelve pages because she needs to show due diligence to the board, and because a thin file looks negligent. Marcus generates twelve pages because he needs to show Dana he did the work. The reader is absent from the transaction. When no one is actually reading, volume becomes the only currency. The machine offers unlimited pages and charges no labor.

Dana did not want twelve pages. She wanted cover. A thick document is proof of effort if the board asks. The machine does not care about her anxiety. It produces the thickness without the effort, which means she no longer even has the excuse that the work took time. The gap has no floor.

Eli, the vendor rep, understood this before his customers did. At the demo he showed Dana how the tool could draft a forty-page report in minutes while she watched. He never asked what decision the report supported, because that question would slow the sale. He knew the sale didn't depend on usefulness. It depended on the checklist. The software checked every box. Dana bought the tool anyway, and scheduled the rollout.

Marcus could have asked Dana which risk she was actually worried about. He would have gotten a one-sentence answer. He could have written two lines and solved the problem. But asking felt like more work than generating twelve pages. He chose the gap.

You can see the same pattern in your own inbox. The memo repeats meeting notes you already read. The proposal answers every question except the one you asked, and the analysis includes background nobody needs because the template demands it. These are request failures. The machine only lowers the cost of a bad request.

You're not immune. The gap lives in your own requests. When you ask the model for "a comprehensive overview," you're asking for pages, not answers. The machine will oblige. It'll always oblige. That's what makes it dangerous. It has no floor and no ceiling. It'll fill any space.

This means the fix is not a better prompt. A better prompt to a hollow request still produces hollow output. The fix is to shrink the gap. Ask for one decision, not twelve pages. Ask for a single paragraph on risk, not a risk encyclopedia. When the ask and the need are the same size, the machine writes what you actually need.

Drill: Pick three documents you produced this month. For each one, identify the exact decision it was meant to support. If you can't name the decision, the document is gap-filling. Rewrite the brief so it asks for that decision in one page or less, then run it through the same tool. Compare the two outputs.

Reading One Story Twice: A Claim Inventory

Marcus dropped the vendor proposal on the table. It was fifty pages, full color, perfect binding. “I’ve read it,” he said. “They can handle everything.”

Dana flipped to page three and looked at one paragraph, then another. “What did they actually promise? Not the theme. The list.”

Marcus paused because he remembered the narrative, the way the vendor started small and scaled up while adding modules. He couldn’t name a single deliverable they had guaranteed.

“You read it for plot,” Dana said. “Now read it for inventory.”

Every proposal is two documents stacked together. The first is a story where the vendor introduces a problem and offers a solution, then wraps up with a bright future. You read this once and you nod because the story makes sense.

The second document is hidden inside the first, and it’s a list of claims. Some are hard promises with numbers. Others are soft adjectives dressed as facts. The rest are assumptions about your business that you never approved.

You need to read twice because one pass can’t do both jobs. Story mode turns off your skepticism, and you’ll miss the fine print. Inventory mode turns off the narrative, and you’ll miss the flow. Do both.

The second read is slower. You stop at every sentence that does work, and you ask what claim it makes and who pays if it is wrong.

Take a paragraph like this one, which Marcus found on page four.

“Acme Logistics Platform connects to your existing ERP infrastructure to provide real-time visibility across the supply chain. Our cloud-native architecture delivers 99.99% uptime, and the familiar interface cuts training time by up to forty percent.”

First pass, it sounds fine. They connect to your system, you see things faster, it stays up, and training shrinks.

Second pass, it is four claims hiding in two sentences.

Claim one is integration. They say their platform connects to your existing ERP, which assumes you run an ERP and that theirs talks to yours. It also promises the connection will work. One sentence yields two assumptions and one promise.

Claim two is speed. Real-time visibility sounds like engineering, but it is not. Real time means five seconds in one room and five hours in another, so ask them which one they mean. They also assume your supply chain is one visible thing when it might be twelve subsidiaries and a warehouse that still runs on spreadsheets.

Claim three is reliability. Ninety-nine point nine nine percent uptime looks like math, and it's a hard promise until you ask about the time period and the measuring tool. Ask too whether scheduled maintenance counts. The number is precise, but they left out the context.

Claim four is training. They call the interface familiar even though they have never watched your people work. Familiar is a claim about your employees made by a stranger. The forty percent reduction builds on that guess, and if the interface is strange, the forty percent vanishes.

Marcus marked the paragraph with two colors. Yellow meant hard promises that needed contract language, and pink meant assumptions about his company that he'd never approved. By page six he had stopped counting and went to get coffee.

This is the claim inventory. You're not deciding if the vendor is right. You're simply writing down every bet they're asking you to cover.

You need three buckets.

Hard promises are sentences that carry numbers and mention dates or deliverables. They belong in the contract. If the proposal says "goes live in ninety days," that's a hard promise, and if it says "reduces error rates by half," that's hard too. These are easy to spot, and they're also the easiest to forget during the story read.

Soft promises are adjectives pretending to be specifications. Easy. Fast. Flexible. User-friendly. These words don't commit the vendor to anything. They commit you to hoping. Mark them in a separate list. Before you sign, turn them into testable requirements or remove them from the deal.

Assumed facts are claims the vendor makes about your business. Your staff are ready for change. Your data is clean. Your current system is standard. Your compliance requirements are typical. They write these as if they are obvious. They're not obvious. They're risks you're being asked to swallow.

Once you have the three lists, you work backward. Send the soft promises back to the vendor and ask them to rewrite each one as a testable requirement. User-friendly becomes a specific test. New hires complete standard tasks without help desk calls within two days. Send the assumed facts back too. Ask which discovery call they learned that on, or whether they are guessing. Hard promises stay on your desk. You'll compare them to the contract line by line.

That is the second read. It's not fun. It feels like breaking a toy to see where the batteries go. But the story the vendor told will not show up on day one. What shows up is the inventory.

Find the last vendor email or proposal in your inbox and read it once for story. Then go through it again with a pen or a highlighter. Mark every hard promise in one color and every soft adjective in another. Mark every sentence that assumes something about your business in a third. When you are done, count how many of each you found. If the soft and assumed lists are longer than the hard list, you do not have a proposal. You've got a brochure. Don't sign a brochure.

Send the vendor the soft and assumed lists. Ask them to turn the soft claims into testable requirements and to justify the assumed ones. Don't rewrite it for them. Let them do the work. The quality of what they send back is your first real deliverable.

PART II

Building the Ten-Minute Habit

Ten Minutes, Same Time, Before the Child

Jenna pours coffee at 5:52. The baby monitor on the counter stays dark. She sits at the kitchen table, sets a physical timer for ten minutes, and opens a plain notebook. She does not check her phone. She does not load the dishwasher. The timer clicks, and she writes.

She writes one paragraph about work waiting at ClearThread. On bad days it is one sentence naming the next step. No one else reads it. No one else can touch it.

Six months ago she tried a different system. She waited until Rosa went to bed at eight. Then she opened her laptop on the couch and promised herself a full hour. She woke up at ten with the screen glowing and her neck on the armrest. The laptop had done nothing for two hours. She had done nothing for two hours. Rosa cried at nine the next night. The hour vanished. Jenna tried again and got the same result.

Waiting for free time is a negotiation you lose every time. The evening hour looks ample at noon. By eight it has shrunk to a chore. By eight thirty it is a nap.

Ten minutes at the same time every day is different. It is too small to shrink. You cannot talk yourself out of it the way you talk yourself out of sixty.

Once Rosa calls for her, the day shifts to reaction. Breakfast, backpack, shoes, the drive to daycare. Those tasks expand to fill every opening. Jenna does not try to write then. She already wrote her paragraph.

She does not try to solve the whole project in ten minutes. She uses a standing prompt. She writes the next smallest step at the top of the page. Then she fills the rest of the page with the obstacle.

The timer is a physical one, shaped like a tomato, from the hardware store. Her phone stays in the bedroom. A phone is a portal to other people's urgencies, and urgencies do not respect ten-minute treaties.

After a week the notebook shows a pattern. Each entry is one step and one obstacle. "Call vendor" and "I do not have the direct number." "Review pricing" and "I need Tom's spreadsheet." The writing is ugly. It does not need to be clean. It only needs to stay on the page.

After a month she had not finished the project. She had thirty paragraphs. Each one named a single clear action. When a real hour finally opened on a

Thursday afternoon, she knew exactly what to do with it. She sent one email and closed the loop.

This is where people get stuck. They hear ten minutes and think it is too small to count. So they plan for a ninety-minute block instead, achieve zero, and carry the failure for days. The ninety-minute block is a fantasy they can postpone without evidence. Ten minutes is a floor you hit even when you are sick or traveling.

The same time matters more than the same duration. Jenna started at six on a Monday. By the third week she woke at 5:48 without the alarm. The alarm sat unused on the dresser.

She writes on weekends too. Rosa wakes later, so the time shifts, but the rule holds. Saturday morning, Sunday morning. The notebook gets two more entries. The habit does not check the calendar.

Before the child means before the world votes on your priorities. No one asks anything of her at 5:52. The news is not demanding attention. The day has not yet begun its sales pitch.

I have tried the evening version. The laptop always falls asleep before I do. Morning is the only meeting that stays on the calendar.

Jenna has one hard rule. The phone stays in the bedroom until the timer rings. The notebook is the first thing she touches. Paper does not offer notifications. It only takes ink.

Her partner, Tom, knows the signal. If the kitchen light is on and the timer is ticking, he stays out. He pours his own coffee later. That small boundary costs him nothing. Jenna gets her ten minutes.

Missing a day does not ruin the habit. Waiting until evening to make it up does. When Jenna skips the morning, she does not write at lunch. She loses one paragraph. The project stays on the shelf another day.

Here is how the ten minutes work in practice. Last Tuesday Jenna wrote at the top of the page: "Vendor contract: email Dana the liability clause by noon." The writing took three minutes. She spent the next four minutes listing what Dana would need to know. Sending the email took two minutes after breakfast, once Rosa was in the car seat. Dana replied by three. The issue closed. Without the morning paragraph, the email would have sat in her mental queue for a week. It had sat there for ten days before.

You do not need a child for this to apply. You need a boundary that the world crosses predictably. For Jenna it is Rosa. For you it might be the first meeting, the commute, or the moment your inbox reloads. The principle is identical. Anchor the ten minutes before the crossing, not after.

Do not make the ritual fancy. A notebook, a timer, and a prompt you can answer in one sentence make up the kit. Buy the notebook once. Use it tomorrow.

Tonight, before you sleep, put a notebook and a pen where you will sit in the morning. Set a physical timer for ten minutes. Pick one question you will answer the moment you sit down. Do it before you open a screen, before you speak to anyone, before the day asks anything of you. Write until the timer rings. Then stop. Tomorrow, do it again at the same time.

The Four Passes

The Four Passes

Jonas set the draft aside and sent Maya the file. He called it done. Twenty minutes later she dropped a hard copy on his desk. Four ink colors, none of them black. Red in the margins, blue on the leads, green along the sources, orange at the closer. Jonas looked up. “I thought this was clean,” he said. Maya shrugged. “It was. On the first pass. You stopped there.”

The temptation is to stop there. You read what you just wrote, fix the typos, and call the chapter finished. That’s one pass, and one pass only finds the mistakes you were already hunting. The structural hole, the bad evidence, the sentence that runs to forty words — they stay invisible because you read with your intent instead of with your eyes.

Clean copy needs four. Each pass carries one job and one lens. If you mix them, you will miss what each is built to catch. Structure on Monday, evidence on Tuesday, voice on Wednesday, gates on Thursday — one pass per day, inside the same ten-minute sitting chapter four gave you. A longer draft takes more sittings, not longer ones. That sounds slow. It’s faster than publishing a chapter that fails review.

Pass one is structure. Do not touch the commas. Do not fact-check the dates. Ask only whether the skeleton holds. Does the chapter open with a scene or a case? Does it teach one idea, or does it smuggle in two? Does it end with a drill, or does it drift into a summary? Check for lists where prose would carry the idea better. Check for paragraphs that are all the same length, which reads as a tone problem before it is a formatting problem. If the frame is bent, fix the frame.

Pass two is evidence. Pull every story, number, quote, and threshold. Ask which of the three routes it takes. If it’s real, is it cited? If it’s invented, is it labeled as a composite? If you stripped the names, would it still claim something checkable? Watch for false precision. A rounded threshold reads as a rule of thumb. An exact count attached to a composite reads as a lie. Mark anything uncertain.

Pass three is voice. Read it aloud. Not in your head. Aloud. You’ll hear the sentence that runs out of breath and the filler phrase that cushions every point. Count contractions. They should land unevenly, not like a metronome. Search for the banned words. Search for the paragraph endings that explain themselves. If a sentence needs three em dashes, break it in half. If you hear yourself hedging, cut it.

Pass four is the gate. This is where you count what can be counted. Em dashes per thousand words. Average sentence length. Instances of the antithesis or the terminal appositive. The counts are blunt. If the budget is four em dashes and you have nine, you have a decision to make. Same for the sentence-length ceiling. Don't eyeball it. Use a plug-in or a manual count. Numbers promised on the cover get checked here. If the chapter claims a forty-one-set sample, you verify you counted forty-one sets.

The order matters. I've wasted an hour polishing a paragraph I later deleted because the structure was wrong. Structure first, because evidence and voice sit inside it. Evidence second, because voice polish on a bad source buys you nothing. Voice third, because you should tune what you are keeping. Gates last, because arithmetic is ruthless and you do not want to gate-check a paragraph you will later cut.

Jonas ran his next chapter this way. The structural pass found two ideas wrestling in the same section. The evidence pass caught a quote he had attributed to a source he never interviewed. The voice pass killed four filler openers. The gate pass flagged six em dashes in two thousand words. He fixed them all. Maya marked the new printout in black ink for the first time.

Drill. Take a draft you have already called finished. Run it the way this book promises: four ten-minute sittings, one pass per day — structure, then evidence, then voice, then gates. Use a different colored pen for each. Do not combine them, and do not stretch a sitting. If the draft outruns the ten minutes, give the pass a second day rather than a longer clock. Record what you find: a structural split, an unsourced number, a banned phrase, a broken budget. Fix nothing until you've marked all four passes. Then fix only what moves the chapter. Count what remains. If the numbers still breach the gate, run a fifth pass. There is also a deadline version: block out one hour and run all four passes back to back. That hour is an optional deep clean for a day the calendar forces on you, not the daily habit. The habit stays ten minutes.

A Timed Run, Minute by Minute

Chapter 6

A Timed Run, Minute by Minute

Marcus taps the timer and lays his phone face-down on the table.

Ten minutes. That's the whole budget for a first pass. He has learned that a longer pass only produces a tired guess.

The proposal in front of him is twelve pages, bound with a plastic clip. He does not start on page one. Page one is where the vendor tells him how much they care.

Minute one. He counts the pages and flips to the back. He wants the price and the term. He checks the name and title of whoever signed. The price matches the oral quote. The term is thirty-six months. The signatory holds a title he has never heard of. He circles it. He will check later whether an officer with that title can sign contracts for the company.

Minute two. He reads the opening paragraph. The proposal opens with a promise. This one says the system will improve daily operations and reduce downtime. It calls the approach a unique method. Marcus does not underline any of it. The opening promises are not warranties. He moves to the scope section.

Minute three. The scope lists fourteen items. He checks them against the pricing table in the back. Four of the items carry no price. That's the first flag. He writes *four orphans* on his pad. If they cannot price it, they cannot deliver it.

Minute four. He flips to the liability section. The text is dense. He reads only the first sentence of each paragraph. One paragraph limits damages to the fees paid in the twelve months before the problem. That's a low cap. Second flag. A year of fees would not cover a month of downtime.

Minute five. His phone buzzes against the table. He does not check it. The timer keeps running. He finds himself reading the same sentence about insurance coverage twice. The sentence runs for five lines and contains three nested exceptions. His mind wants to wander because the text is hard. He marks the page and moves on.

Minute six. He checks the references. There are none. Not a named client, not a case study. The document says details are available upon request. Third flag. He writes *no refs*. A vendor with happy customers names them.

Minute seven. He goes back to the dense insurance paragraph. The paragraph is a single block of text with no subheadings. The vendor says they are not liable for delays caused by outside logistics problems. He notes the phrase. Later he will ask who counts as an outside logistics provider. For now, he knows the vendor can blame the shipping company they chose themselves.

Minute eight. He looks at the payment terms. Net thirty. Standard. But the appendix mentions a setup fee due on signing. That fee was not in the oral quote. Fourth flag. He circles the number. A fee that appears only in the fine print is a fee they hoped he would not notice.

Minute nine. He has four flags and he has handled twelve pages. He does not touch the cover letter or the company history. He flips to his notepad and writes one line. Price mismatch. Liability cap. Orphan items. No refs.

Minute ten. The timer rings. He closes the folder and pushes it to the corner of the table. The first pass is done.

Dana taught Marcus this drill. She runs it on board decks before she lets herself form an opinion. I learned it from a contract I once read the slow way — a long afternoon before I finally hit the termination clause buried deep in the annexes. A ten-minute pass would have caught it. The clock protects you. Without the timer, Marcus would still be on page four, convincing himself he was being thorough.

The goal of a first pass is not to understand every word. The goal is to decide if the document deserves a second pass. I have run this drill on maybe thirty documents. Most revealed a hole I could drive through. You have saved yourself an hour.

There's a trick to the timer. It must be visible and it must be real. A mental note to stop in ten minutes is not a timer. You will negotiate with yourself at minute eight. You will decide that this paragraph deserves more time. The paragraph always deserves more time. Give it more time and you're no longer doing a first pass. You're doing a slow read that happens to have started recently.

Here is how to set up your own run.

Pick a document you have been avoiding. Decide on a budget. The sitting is ten minutes — that number does not move. What moves is coverage: a short memo fits whole; a hundred-page contract gets carved ten pages at a sitting, more sittings rather than a longer one. The point is to set a ceiling.

Before you start, name your targets. Marcus looks for the price and the liability. He checks whether the vendor has real references. Your targets depend on your work. Maybe you need deadlines. Maybe you need to see who decides what when the plan slips. Write your targets on a card and look only for those. If the text does not speak to your targets, you skip it. The vendor's company history does not speak to Marcus's targets. Neither does their photo of the warehouse.

Keep your notes on a separate pad, not in the margins. Margins tempt you to fix the language. You're not editing. You're deciding.

Set a hard timer. Put the phone on airplane mode. When the timer rings, you stop. Even if you are mid-sentence. The drill only works if the endpoint is real.

If you finish before the timer, do not fill the time with more reading. Close the file and write your summary. One paragraph. What they want, and what would make you say no. If you cannot name the objection yet, you have your next question.

Schedule the second pass only for documents that survive the first.

Now run it.

Take a document you have been putting off. Set a timer for ten minutes and start it. Read for the frame, not the details. When the timer sounds, close the file. On a blank page, write two sentences. What they are asking for, and the one thing that would make you say no. If you cannot name that thing yet, you have your next question.

PART III

Spotting the Specific Hallucination

Numbers That Sound Right

Sam slides the report across the conference table, and Dana stops at the second page. She reads the line aloud.

"The rollout will free up four hundred and thirty-seven hours per quarter," she says, looking up. She asks where it came from.

Sam shifts in his chair because he started with four hundred, and it felt too round. He added thirty-seven to make it look like he had done the math. He had only guessed.

This is how numbers fool us. A precise figure sounds like a measured fact, but look closer and you see a guess in costume.

Readers treat specificity as proof, so show them 437 and they picture a stopwatch and a spreadsheet. Show them "about four hundred" and they picture you guessing.

In business writing this happens constantly, with projected savings arriving at $847,500 and customer satisfaction estimated at 87.3 percent. Such numbers wear the outfit of hard data, but they did not come from measurement.

I've done this myself, because early in my career I wrote that a training program would cut errors by 23.5 percent. I liked the decimal because it made me sound like I'd run a regression, though I'd only asked one supervisor what he thought.

The rule is simple, and it is strict: a number must carry its origin or its uncertainty. If you counted, say so, and if you estimated, say that too. Don't let the shape of the number do the talking for you.

Marcus in finance sees this every week and calls them calculator numbers. They show up with decimal places that imply a chain of division, though the input was only a thumb in the wind. A writer who admits the guess earns his trust.

Sam rewrites the memo and starts with the old claim. The new intake process will save 437 hours per quarter and cut processing costs by 18.7 percent. He can't defend any of that. He has a hunch that the team wastes time on duplicate entry, so he asked two people for their guesses.

He replaces the fiction with honesty, writing that based on estimates from two team leads, streamlining intake could save roughly four hundred hours per quarter. He adds the truth: they haven't timed the current workflow, so the reader should treat that as a placeholder.

If he later runs a pilot, he can tighten the number. In a test group of six, the new process saved an average of thirty-seven hours per person each month. Scaled to the full team of twelve, that is roughly four hundred forty-four hours per quarter. You can see the source, the math, and the remaining guesswork.

When you don't have a number, you have three honest moves. You can use a range, because "between three hundred and five hundred hours" signals uncertainty without pretending to be a calculation. You can name the source of the guess, since a line like "Two team leads estimated…" tells the reader whom to ask. Or you can drop the number entirely, because sometimes the right answer is simply "a significant amount of time, which we haven't yet measured."

The decimal is the worst offender, because 18.7 percent is not more honest than 18 percent. It's just harder to admit you made it up. If your model is rough, your output should show the same roughness.

I still catch myself wanting to add precision I didn't earn, like writing that a project took 142 hours instead of "about three weeks." The single number looks disciplined, but it's not. It's a habit that borrows credibility it never earned.

Dana doesn't want theater, because she wants to know whether she can trust the figure enough to act on it. You earn that trust by showing its source.

Pull the last three memos or reports you wrote that contain a number, and ask each one where it came from. Cite the spreadsheet if that's its home, and rewrite the figure as an estimate if it came from your head, naming the basis. Round false precision to the size of your actual uncertainty. Read the sentences aloud. If a number sounds more certain than it is, fix it before you send.

Names, Quotes and Books That Don't Exist

Marcus typed the closing paragraph at eleven. He needed one more authority to anchor the claim about meeting fatigue. The model gave him a perfect string. Dr. Arthur Blackwell, *The Focus Frontier*, Northfield Press, 2018, page 89. "A team that meets for more than fifteen minutes begins to shed cognitive payload at roughly the rate of a waking dream."

Marcus read it twice. It sounded like Blackwell. It had the academic ring, the italics, the page number. He dropped it in and sent the draft to Dana.

Dana checked it that afternoon. She searched the online catalog. No Arthur Blackwell had written *The Focus Frontier*. She searched the sentence in a book database. Nothing. She tried a scholarly archive. Silence. The quote was a composite dressed in pinstripes.

That's the first thing to know. A fabricated citation doesn't look fabricated. It looks more authoritative than the real thing. It has page numbers, publishers, years, and names you almost recognize. The model isn't lying to you. It's completing the pattern that authority usually takes.

You can't catch this by reading for tone. The tone is exactly why it fools you. The model has read more footnotes than any human alive. It knows that a string like "124 *Policy and Practice Review* 892 (2011)" signals erudition. So it produces the shape and trusts you to fill in the substance.

Here's how to test any citation before you let it travel.

Hunt the exact quote first. Paste a distinctive eight-word string into a search engine with quotation marks. If the only results are your own document, the quote is orphaned. Real quotes leave tracks.

Next, check the container. If the source is a book, find the catalog entry. Search an online library or bookseller database. If the title returns nothing, you've got a ghost. If the book exists, pull a preview or digital copy. Verify the page. Page numbers shift by edition, but the chapter should exist.

Finally, distrust the famous name. A fake quote is safer when it rides a famous name. The more convenient the quote, the more likely it was minted last Tuesday.

I learned this the hard way — the scene that follows compresses a real episode, details softened. Early in my editing work, I let a client cite a paper on cognitive load. The citation had authors, a journal title, a volume number. I assumed someone else had checked it. Weeks later a fact checker called.

The journal was real. The volume was real. The article was not. The client had to issue a correction.

The danger spreads sideways. One fake citation dropped into a report gets copied by another analyst. Soon three decks cite the same nonexistent study. The fiction hardens into institutional memory. You hear it echoed as "research shows" in a meeting. At that point, the lie has a pension plan.

If you need a source and you don't have one, you've got three honest routes. You can go find one that actually says what you claim. You can drop the claim. Or you can flag it as your own summary and take responsibility for the words.

Let me show you what the check looks like in practice. This is a composite, invented for illustration.

Say you're writing about meeting efficiency. Your draft says: "According to Dr. Helena Voss in *The Eight-Hour Myth* (Northfield Press, 2019, p. 44), meetings over fifteen minutes reduce retention by forty percent."

You search "Helena Voss" and "The Eight-Hour Myth." No book. You search the phrase "meetings over fifteen minutes reduce retention." The hits are blog posts that quote the same line, none older than two years. You shorten the search string. Still no academic source. The forty percent is too round. The name is too convenient.

Now rewrite honestly. If you've seen the pattern yourself, say so. "In my experience, meetings past the fifteen-minute mark lose people." If you have a real study, swap it in. If you have neither, cut the claim.

One warning. Don't replace a fake citation with a real one that doesn't quite say what you claimed. It's still wrong. If the real study says meetings over thirty minutes matter, don't write "over fifteen" and cite it anyway.

Your drill. Pull up a document you wrote in the last month. Pick three citations you added because the model suggested them or because they sounded right. Run each through this triage.

Start by searching the exact quote. Does it appear anywhere outside your file and the output that spawned it?

Then verify the container. Can you find the book, the paper, the case, the statute? Does the page match?

Last, ask whether you chose the citation because it supported your point. Or did you settle on the point first and then hunt for any citation that would fit? If you can't confirm it, cut it. Replace it with what you actually know, or with nothing.

Do this now, while the chapter is fresh. The quotes that survive are the ones worth keeping.

Twelve Fun Facts, Checked

Gary clicked to the next slide. Twelve Fun Facts About Communication. Dana leaned back. The first bullet said that ninety-three percent of meaning comes from tone and body language. Gary smiled. He said it's why their platform mattered.

The room nodded. Dana kept her eyes on the screen. She wrote one word on her pad. Source?

After the demo, Dana sent the deck to Marcus in procurement. She'd added a note. Check these. Then tell me how many are real.

Marcus opened the file that afternoon. He'd started with the big number. Ninety-three percent nonverbal. A minute of searching took him to a 1971 study on mixed signals in single words. It covered single words paired with mismatched tone. Subjects were asked to judge which channel carried meaning when words and tone conflicted. It wasn't a study of general conversation. It wasn't even a study of full sentences. The 93% figure was a mangling. Marcus labeled it distorted.

He moved to the second fact. Humans now have an attention span shorter than a goldfish. Eight seconds. He'd followed the citations through three decks and two white papers. The trail ended at a marketing report that cited nothing. He labeled it invented.

The third fact claimed that multitasking cuts productivity by forty percent. He'd found a small lab study with twelve subjects. The slide had turned a narrow finding into a universal law. He'd labeled it stretched.

The fourth fact said that people remember ten percent of what they read but eighty percent of what they do. He'd seen this one a dozen times. It had appeared in training manuals. He'd traced it to a single consultant's worksheet from the 1960s. There wasn't a study behind it. He'd labeled it folklore.

The fifth fact claimed that fifty thousand thoughts run through the human mind each day. He'd checked medical databases. He found a 1970s self-help book that guessed at the number. He'd labeled it invented.

The sixth fact said that seventy-five percent of change initiatives fail. He'd found a single 1995 article that mentioned a range. Over the years the range had become a single percentage. He'd labeled it bent.

The seventh fact stated that customers tell an average of nine people about a bad experience. He'd found a 1980s white paper about complaint handling.

The paper mentioned one study with a small sample. The nine had become an average through repetition. He'd labeled it worn out.

The eighth fact said that visual information is processed sixty thousand times faster than text. He'd laughed out loud. He'd seen this in five vendor decks that year. Nobody could name the lab that measured it. He'd labeled it science fiction.

By five o'clock he'd closed all twelve. Two held up. One was a real census figure. One was a genuine industry average from a named trade association. The rest were garbage. Five were distorted versions of real research. Five had no origin he could find. He'd sent Dana a one-line email. Two for twelve.

Dana replied in minutes. Put it in the contract review.

Marcus walked the findings into Dana's office the next morning. She'd printed the deck. The pages were on her desk. She'd circled every fact in red. She asked him how long the checks had taken. He said under an hour. She asked how much the platform cost. He said six figures. She said that's a good hourly rate.

The vendor called two days later. Gary asked if they were moving forward. Dana said they weren't. She didn't explain why. She didn't need to explain. The proposal had been built on claims no one owned.

I've seen it before. These lists circulate like office colds. Someone adds color to a deck. A number sounds right. Nobody asks who counted. Soon the fact's in every proposal. Then it becomes the premise for a budget line.

I've watched boards approve spending because a slide said researchers found. Nobody asked which researchers.

Marcus ran three checks on each claim. He asked for the named source. If the slide said researchers found, he wanted the name and the paper. If the only citation was another blog post, he stopped. There wasn't a source. Then he checked that the source matched the claim. A real study's often about a narrow condition. The deck turns one lab result into a law of nature. Then he checked whether the source was evidence or just an earlier deck. A report that cites another report that cites a tweet is a mirror.

The process was mechanical. It didn't require special training. It only required the refusal to nod along.

Vendors rarely lie outright. The problem is that nobody in the room demands the footnote. The fact feels right. It matches what people already believe. It gets a pass.

A checked fact takes five minutes. An unchecked fact'll take a budget with it. Check the source, the match, and the rigor. Then move on.

Be clear about which job Marcus was doing that afternoon. Twelve claims in one sitting is the deep sweep, the pass you run when a whole deck lands at once, and it costs what it costs. The daily habit is smaller: ten minutes, the one or two numbers that cross your desk today, checked before they settle in. Run the ten minutes every day and the hour-long sweep stays rare, because the folklore gets caught one fact at a time instead of piling up twelve deep in a vendor deck.

Your turn.

Pull up the last deck you presented or approved. Find every number without a named source. Ask where it came from. Check the original. If the trail ends in air, cut it. Replace what fails with what you can prove. Then send it back. This one is a sweep, not your daily ten minutes. Give it its own slot, once, and let the habit keep it from being needed again.

PART IV

Managing the Error Once Found

The Correction Line

The Correction Line

Walt closed his laptop at six. His draft was done. Then he saw the word count. Twelve hundred words sat below the ceiling. He opened the document again. A draft that thin looked lazy. He started adding.

He wrote transitions. He wrote clarifications. He dropped in a second example, then a third. By eight he had hit the target. He told himself the extra material made the argument stronger. He sent it to Priya.

Priya sent it back the next morning. The file bled red. She had cut the new examples, the transitions, and a paragraph he had thought was essential. She left a one-line note. “Please tighten.”

Walt stared at the screen. He felt the draft was underexplained. He felt the reader needed more hand-holding. For every red line Priya had drawn, he wrote two new sentences saying the same thing in softer words. He added context. He added qualifiers. By noon the draft was four hundred words heavier than when he had started.

Walt told Priya he had tightened it. The draft was four hundred words heavier. She did not reply.

This is not correction. It’s drafting in a different font. Walt thought he was refining. He was actually protecting his own discomfort with white space. Once you finish a draft, you’ve crossed the correction line. The work on the other side is different. You don’t build. You strip.

Look at the paragraph Walt added.

“Prior to finalization, it is worth noting that the proposed integration will leverage existing infrastructure to deliver a seamless end-to-end workflow. This is a game-changer for stakeholders, and it empowers teams to unlock new efficiencies while reducing overhead.”

Priya cut the whole block. She replaced it with one sentence. “The integration uses existing pipes. Overhead drops and teams move faster.”

The forty-seven words carried no new cargo. They repeated what the first sentence said. They only made noise. Noise is hard to spot in your own draft because it sounds like diligence. It sounds like thoroughness. Read the passage aloud. If your own voice starts to drone, you’ve found noise.

You finish a draft. You open it the next day. A paragraph looks bare. You dress it up. You add a throat-clearing phrase before the plain fact. You add a second example where one was enough. You’re still drafting.

Once you cross the correction line, your job changes. You find the sentences that do work. You delete the ones that don't. Work is a single fact, a single instruction, or a single step. Anything that frames the fact, repeats it, or warms the reader up to it is not work. It's packaging. Throw away the packaging.

Keep three checks in front of you during a correction pass.

First, you may delete or swap. You can't add. If a sentence is missing a fact, don't write the fact. Mark the gap and keep cutting. A hole is honest. Leave it empty. Come back in a day. The hole that still bothers you deserves a fact.

Second, keep the ban list in view. Tired writers reach for familiar padding. The phrases you cut in chapter three will try to sneak back in while you correct chapter ten. You'll reach for the same throat-clearing phrases at three in the afternoon because your brain wants a shortcut. Treat them like weeds.

Third, weigh the draft before and after. The result must be lighter. If the count goes up, you reheated the draft. Run another pass.

Walt tried again. He opened a different section. The paragraph read: "In order to ensure robust scalability, we recommend a phased approach that leverages modular architecture prior to full deployment."

He looked at his three checks. He wasn't allowed to add. He cut "In order to." He cut "robust." He swapped "leverages modular architecture" for "uses modules." He replaced "prior to full deployment" with "before full rollout." He lost half the words. The sentence said the same thing. It looked wrong to him. Lean paragraphs always do. They look wrong because you're not the reader. You're the surgeon. He moved to the next paragraph.

Drill

Take a finished draft. Note the word count. This is a correction pass, and a correction pass is not the daily habit. The ten minutes from chapter four stays its own size and its own hour; this drill runs when a draft crosses the line, so give it a separate slot and a bigger clock. Set a timer for twenty minutes. During that time, you may only delete or replace words with shorter ones. You can't add sentences, examples, or qualifiers. When the timer stops, check the count. If the draft isn't lighter, run it again tomorrow. Keep the surplus words off.

When You Can't Find Out

When You Can't Find Out

Northfield's rep had just emailed the capacity chart again. Jen opened the attachment and saw the same bar graph she had already received twice, as if resending would change the ink. Twenty thousand units a month. Peak. Sustained. Her engineering lead, Raj, needed eight. The margin looked thick. But the one Northfield customer who had agreed to talk to her had left three voice mails promising a callback. None came. The facility was in Malaysia. A site visit would cost twelve thousand dollars and burn ten days. Raj needed the purchase order by Friday. Jen sat with her cursor blinking over the draft to her boss, unable to choose between "I think we're good" and "I still don't know."

You have a file like this on your desk. The question is not academic. Something real depends on it. And the information you need is either hidden, expensive, or late.

This is the point where most people stall. They send another email. They schedule another call. They look for a paper or a person that will make the doubt disappear. You tell yourself you're doing due diligence. What you're actually doing is waiting for certainty that won't arrive.

I've learned to call this the verification trap. You treat the missing fact as the bottleneck. You believe the decision cannot move until the fact is found. It feels responsible. It looks like diligence. But diligence has a deliverable: a decision. Searching past the point of return is just a slow way of saying no, except you don't own the no. The project dies of boredom and nobody learns why.

There's a way out. It's not to search harder. It's to admit you cannot find out, then change the question.

When you cannot verify a claim, shift from asking "Is this true?" to asking "What do I do if it's false?" The first question demands proof. The second demands a plan. You can make a plan with missing information. You can't make proof appear on command.

Jen didn't need to know Northfield's true capacity. She needed to know whether to sign a purchase order that locked her in for a year. Those are different decisions. Once she separated them, the path cleared.

She grabbed a notepad and wrote the decision at the top. Should we single-source eight thousand units a month to Northfield? Then she drew a line. Under it she wrote the threshold: 8,000 units minimum sustained. She stared at the number. She didn't need Northfield's ceiling. She needed to know the floor.

Raj stopped by her cube that afternoon. He asked if she was ready to cut the PO. She told him the reference was still outstanding. He shrugged. "So call them again." She had called them twice. They weren't going to talk, and she wasn't going to Malaysia. She had to decide anyway.

What she wrote down was the action question, not the research question. "Should I single-source eight thousand units a month to Northfield?" is something she can answer. "What is Northfield's maximum capacity?" is not. It is a data point, not a destination.

Next, find the threshold that would change your mind. Ask what number would flip you from yes to no. Raj needed eight. Jen would feel safe with a supplier rated for twelve. She didn't need twenty. She needed to know if the floor was above eight. That threshold is your real search target. It is much smaller than total certainty.

Most people ask for the whole file because they think more data reduces risk. It does not. More data reduces anxiety. Risk lives in the gap between what you know and what you decide. The threshold closes that gap with the smallest possible fact.

In Jen's case, even the threshold was opaque. She had Northfield's word, a stale case study, and silence. She couldn't reach the threshold, so she stopped trying to reach it.

If you cannot hit the threshold with confidence, build a hedge instead of a proof. A hedge is a cheap way to survive being wrong. Jen called a local machine shop she had used three years earlier. Their quote was high, but they could start in forty-eight hours. That speed was the hedge. She priced the extra cost into the contract. If Northfield stumbled, she would pay more for a few months. She wouldn't stop production.

Then she added a contractual tripwire. If Northfield missed two consecutive delivery dates by more than ten percent, the contract allowed her to release thirty percent of the order to the local shop without penalty. She didn't need to know if Northfield would fail.

This is how you operate when facts are underground.

There is one more tool. Set a kill date for the search itself. Pick a day and time when the open question officially closes. At that moment, you commit to the best available decision with the best available hedge. If the information shows up late, it waits for the next cycle. The kill date prevents the elegant inertia of permanent research.

In my experience, teams can spend six weeks trying to verify a single software claim. The vendor documentation is vague. The engineers send questionnaires. Legal sends clarifications. The client moves on. The team is left with a full inbox and an empty contract.

A kill date at week two would have forced a different outcome. Either they would have accepted the ambiguity and added an audit clause, or they would have eliminated the vendor and chosen another.

You can already see the pattern. The original question tempts you to treat the missing fact as the problem. The real problem is the unmanaged risk beneath the decision.

Try the shift on a question that is sitting in your inbox right now. Maybe it's a hire whose references never call back. Maybe it's a market size you can't afford to study. Maybe it's a supplier like Northfield.

Write the decision you actually have to make.

Under that, write the threshold. What is the minimum number, trait, or condition that would let you say yes? Be specific. "Good culture fit" is not a threshold. "Can ship ten thousand units by March" is.

Under that, write the hedge. If the answer turns out to be worse than the threshold, what can you do? Who can you call? What clause can you add? What budget can you reserve?

Then write the kill date. Pick a real day. Add it to your calendar. When it arrives, the search is over. You're now in execution mode.

Jen sent the purchase order on Thursday. She included the tripwire clause and reserved the local shop's capacity for the first quarter. She never did get the reference call. The production line started on time. Her boss never asked about the reference. He asked whether the parts arrived.

Drill

Pick one open question that is blocking a decision on your desk right now.

In fewer than fifty words, write the actual decision you need to make. Not the missing fact, the action.

Next, write the threshold. What is the specific number, condition, or trait that would let you move forward with confidence?

Then list one hedge. If you're wrong, what is the cheapest way to survive it?

Finally, set a kill date. Put it in your calendar. When that date arrives, commit to the decision with the hedge.

Do this now for that one open file. Then close the file and send the email.

Bedtime, Interrupted: A Correction in Full

Your phone buzzes against the nightstand. You ignore it. It buzzes again. You pick up. Sam's voice is two ticks faster than usual. The bid for Fenton Group went out at six, and he just opened the attachment. The pricing matrix is last year's template. Every number downstream is wrong.

You stand in the hallway in socks. The proposal's due by nine in the morning. You picture opening your laptop, swapping the grand total, and crawling back under the covers. That fix would bury the error where no one sees it until tomorrow afternoon.

You know the difference between a typo and a structural error. A typo embarrasses you for a moment. A structural error makes you wrong in every direction, and it does not stay in one cell. At bedtime, both look like small problems. The difference only matters in the morning.

This is the bedtime correction. The hour makes everything feel urgent. Urgency tricks you into treating a foundation error like a surface scratch. A wrong template is not a typo. It's a seed. It puts down roots in every cell that pulls from it. By morning it will have spread through the labor rates and burden formulas until it reaches the executive summary table.

If you change only the summary table, the line items still carry old overhead rates. Adjust the line items and miss the burden schedules, and the footnotes contradict the body. Fix the footnotes but forget the Appendix B rate card, and Ruth at Fenton has two documents that don't match. Each partial repair creates a new seam. The document becomes a trap. Ruth will find it before lunch.

The morning after a patch feels fine. You got four hours of sleep. You tell yourself you fixed it. By noon Ruth calls. The line item for labor escalation does not match the summary table she printed. She forwards your email to her director. Now you are explaining two versions of the same error instead of sending one clean file.

You need a correction in full. Stop the file. Recall what you sent. Rebuild from the corrupted layer. For a proposal, the corrupted layer is the template. Start there. Pull the current rates from the shared drive and drop them into a clean sheet. Then repopulate every cell that depends on the new math. Check the sums. Check the cross-references. Check the escalation clauses in the narrative against the cells in the back. Save the file with a new version number. Then send it with one line saying you replaced an early draft.

Sam can help. Give him the original template from the shared drive. Ask him to flag every section that pulls from the matrix. While he works the bones, you work the narrative. Split the labor. Don't split the file into patchwork. One person holds the master until it is whole. If you both touch it, you'll bury the document in conflicting edits.

Never review a midnight correction alone. Your brain will see what you meant to type, not what the screen shows. Sam should open the file cold and read the first page without your voice in his ear. If he pauses over a number, you have missed something. That pause catches the second error.

The client will not love a late email. They'll love a wrong bid even less. A corrected file at seven in the morning beats a broken file at six. Do not apologize for the delay. Apologize for the version they should not have seen. Then stop talking. An explanation invites them to compare the two files. You do not want a client doing forensic analysis on your mistake.

By 11:30 your laptop is hot on the kitchen table. Sam has rebuilt the labor tables. You have checked the escalation clauses against the new burden schedule. The new file is clean from the first page to the signature block. You write Ruth at Fenton Group. "Please disregard the previous send. Attached is the corrected proposal." Version numbers matter. If you send the same filename, Ruth will open the wrong one by accident. Name it clearly. "V2" in the filename is enough to signal that the first file was dead on arrival. You don't explain why. An explanation invites questions. You hit send.

Then you sleep. You won't sleep well, but you'll sleep honestly. The correction took an extra hour. It saved you from a meeting where Ruth asks why your firm miscalculated benefits by eleven percent. There is no good answer to that question.

In my experience, the halfway fix is the real risk. The error you catch is rarely the only one. When a bad template or a bad assumption enters a document, it leaves fingerprints everywhere. Fixing the headline and hoping the body follows is like repainting the hood when the frame's bent. The car looks fine until the first turn.

The halfway fix is also a habit. Once you patch one document at midnight, the next error feels easier to patch. Each shortcut trains you to trust sleep over rigor. I have sent drafts with band-aids on top of band-aids. That is how eleven percent becomes company policy.

The next time a call interrupts your evening, ask one question before you open your laptop. Does the error sit on top, or does it run through the middle? If it runs through the middle, close the screen until morning. Then rebuild the whole thing in daylight. Partial repairs are for surface problems. Foundation problems need you to rebuild from the source.

The seam you find in daylight was there at midnight.

Drill. Pull the last important document you sent. Pick the longest section. Check whether the numbers in the body match the numbers in the summary. If they do not, you have found a seam. Do not patch it. Trace it back to the source cell or paragraph. Find the assumption that feeds it. Fix it there. Then check every other place that same source appears. That is a correction in full. Run it now.

PART V

Systematizing Accuracy Across Teams

The Household Error Log

Avery turned the knob to medium and the oatmeal boiled over anyway. The burner on the left ran hot. Jordan appeared in the doorway with a mug of coffee and stopped. The pot hissed. It was the third Tuesday in a row.

They didn't fight about oatmeal. They fought about attention. Jordan said the stove was broken. Avery said she had watched the pot. Both statements were true, and neither solved the problem. The mess stayed on the stove while they left for work.

You have a version of this. A small failure repeats. You blame the appliance, or you blame the person standing next to it. Nothing changes because the error lives in memory, and memory fades by Thursday.

Avery ran operations at a small clinic across town. Her office kept a running log of every missed delivery, tripped breaker, and mislabeled vial. The rule at work was simple. Log the event, the conditions, and the fix. No names. She printed a blank table at the clinic and taped it inside her own kitchen cabinet when she got home.

This is the household error log. It treats domestic failures like system bugs. The form is spare. You write the failure, the context, and the change to test. You write it while the mistake is still warm.

In my experience, the hardest part isn't the review. It's stopping to write the entry before you wipe up the mess.

The hard part is the wording. You describe the failure without assigning a villain. Not "Jordan overloads the washer" but "Washer stops mid-cycle with unbalanced load." Not "Avery forgets the mail key" but "Mail key is not in the bowl when shoes go on." The person is not the bug. Write about the system.

Here is how Avery filled the first month.

The oatmeal entry went in on Tuesday morning. The knob marking was off and the lid sat off the pot. The fix was tape on the knob, burner right for oats, lid stays off. She tried it the next morning. The oatmeal cooked evenly. She taped the knob before the pot cooled.

The trash bag split on Thursday evening in the hallway. The bag was store-brand and the wet coffee grounds sat at the bottom. She bought thicker bags. They worked, and she didn't need to double-bag.

The bathroom ran out of toilet paper on Saturday night. The backup rolls lived in the hall closet behind the laundry basket. Jordan moved a spare roll to the tank lid. The next Saturday, the spare was still there. They closed the entry.

The car battery died on Monday morning in the garage. The dome light was left on, and this model had no audible warning. Avery wrote the fix as a sticky note on the driver visor to check the light before locking. The note lasted three days before it fell. She logged a revision. The adhesive was weak in winter cold. The fix became a magnetic clip on the visor. The clip held.

Three of the four fixes cost nothing. The tape took ten seconds. The sticky note solved the car issue before the next snowfall. The toilet paper moved six inches. Only the trash bags required money. Even that was a product swap.

Notice what the log is not. It's not a to-do list. To-do lists nag. They sit there and judge. The error log waits for a pattern. If an entry happens once and never repeats, you leave it open and forget it. If it repeats, the log shows you the shape. You don't chase every open item. You chase the ones that visit more than twice.

The value compounds at the review. Once a week, usually Sunday evening, Avery read the last seven entries aloud while Jordan wiped the counter. Patterns stood out when the page held more than three incidents. The washer appeared twice. The mailbox key was missing on three separate days. The review took twelve minutes. They picked one fix to install before Monday. They did not try to fix them all.

You will want to turn the log into a record of wrongs. Resist. The goal isn't to document who failed. The goal is to make failure expensive for the house and cheap for the people inside it. Tape on the knob means Avery no longer has to think about oatmeal on Tuesday mornings. The system holds the fix.

Jordan started logging too. His first entry was about the coffee grinder. The blade jammed on dark roast beans because the motor housing shifted when the machine was pushed back against the backsplash. He added a rubber foot from the junk drawer. The grinder has not jammed since.

After eight weeks, the log held thirty-two entries. Sixteen were marked closed. The rest clustered around two themes. Morning exits and grocery storage. Avery could see the shape only because the log held enough data. One more week and the pattern would fill a second page. That is when she bought the shelf bins and moved the granola bars to the left of the refrigerator. Morning friction dropped.

You can build this in any format. A notebook on the counter. A note on your phone. A shared document that both of you can reach. The container doesn't matter. Pick one tonight and write the first entry from today.

Log the failure while it is fresh. Review it while it is still legible. Fix one thing before the list becomes a catalog of complaints. Leave fault for the living room. The log holds fixes.

Drill: Run your own log for seven days.

Pick a place. Tape a sheet inside a cabinet or start a note on your phone. Log every failure as it happens. Use three fields: Failure, Context, Fix to try. Keep each entry to one sentence. No names. If you live alone, write in the second person for yourself. "I left the windows down" becomes "Windows were down during the storm."

At the end of the week, read the entries aloud. Circle any item that appears twice. Pick the one that would remove the most friction. Install the fix before the next week starts.

Run this for one week. The log will show you where your house is actually broken.

Homework Bots, Read-Alouds and the Handover

Homework Bots, Read-Alouds and the Handover

Sam Chen had half an hour. He pasted the brief into the prompt, copied the bot's reply, skimmed it for typos, and hit send. He didn't read it aloud. He missed the line that stretched the warranty to a decade, and the clause that promised onsite repairs no one had approved.

Two days later the client forwarded the thread to Vera Ortiz. She walked to Sam's desk with a printout. "Read the third paragraph out loud," she said. Sam did. His face went hot. The promise was absurd. So was the number. A machine had done the homework. Sam had skipped the read-aloud, so the document never changed hands.

AI output looks finished. The grammar's clean and the tone's even. There are no misspellings to catch your eye. It carries the weight of a final draft because it sounds like one. That confidence is the trap.

Your tongue doesn't forgive. When you speak text exactly as written, you trip over swapped numbers. You hear a clause that belongs to a different product line. You notice a phrase that no one in your company actually uses. The bot borrowed it from a training document and dropped it in like a spare part. On the page it looks technical. In your mouth it tastes borrowed.

The read-aloud is a filter, not a performance. You stand, or sit, and say the words in order. You may feel foolish talking to your screen. Do it anyway. The awkwardness is the point. If you're embarrassed to say it, imagine sending it.

The handover is the gap between generation and ownership. The bot does the homework first. It expands bullets and drafts replies. It restructures notes. That's its job. Your job begins after. You read the work aloud and mark every hitch. A missing comma crashes. A double meaning lands. A statistic you never provided sits there like an uninvited guest. You fix them. Then the words become yours. You sign it, send it, or present it. Before that step, the document belongs to the machine. After it, it belongs to you. The read-aloud is the line you draw before you sign.

Sam now keeps a sticky note on his monitor. It says: Aloud, then out. Vera does the same with contract terms. She caught a limitation-of-liability clause last month that limited damages to the annual price of "a reasonable sandwich." The bot had mangled a placeholder. The spelling checker found nothing. The grammar tool flagged nothing. Only Vera's voice caught the absurdity, because she slowed down and the words reached her ears.

The problem is not that the bot makes things up. The problem is that it makes them up in the same confident tone it uses for facts. An invented clause looks like a real clause. A hallucinated date looks like a real date. Smooth writing is not verified writing. Reading aloud does not verify truth, but it forces you to inhabit the text. You have to own each sentence before you can speak it. You can't send a sentence you can't speak.

You don't need a special room. You don't need a script. Take the draft, close the door if you have one, and read from the first word to the last. Don't skip the greeting. Don't skip the boilerplate. The bot hides its worst errors in the places you think are safe. Read the subject line aloud. Read the signature block. If your voice falters, mark the spot. If you rewrite a phrase in your head before your mouth says it, mark that too. Those marks are your real edits. Everything else is formatting.

If the document is long, read the summary aloud. If it is a deck, read the slide titles in sequence. If it is a script for a video, say it to your reflection. The ear catches what the screen hides because the ear hasn't learned to skim AI prose. The ear hears what is actually on the page.

There's no substitute for this step. Saving the draft for later doesn't work. Your eye will make the same assumptions at six pm that it made at noon. Running a grammar checker doesn't work. The grammar is already perfect. Asking the bot to check its own work doesn't work. It will confirm that the text is fluent, which is true and useless. Only your own voice, reading your own words, closes the handover.

Sam's client accepted a corrected letter. The relationship held. But the lesson stayed. The bot had done the draft in under two minutes. The read-aloud would have taken a couple more. Those two minutes bought him a letter worth sending.

DRILL: The two-minute handover

Take the last AI draft you considered finished. Set a timer for two minutes. Read it aloud in full. Mark every place you pause, stumble, wince, or mentally rewrite. Fix those spots. Then ask whether each claim is one you intended, or one the bot assumed. Edit accordingly. When you are done, the document is no longer bot output. It is yours. Send it.

Four Weeks of Passes

Marcus set three pens on the conference table, red for structure, blue for numbers, black for voice. Dana watched him line them up next to the stack of paper. It was a Thursday afternoon at Perimeter Logistics, and their proposal to the transit authority was due in four weeks. Marcus had already printed the current draft, and it weighed almost as much as his laptop. “Last time we used all three at once,” he said. He’d told her about the Birchfield proposal six months ago, and he’d read that draft once, slowly, marking every flaw he saw. By page eight his red and blue ink had blended into purple smears. The staffing section still listed last year’s rates, and he’d missed them because he was busy rewriting a sentence in the opening. Birchfield lost the bid, and the evaluator’s notes said the budget did not match the staffing plan. Marcus could not argue.

Dana picked up the red pen. “Rounds,” Dana said.

“One week per pass,” he said. “Four weeks, four jobs, and each pass gets one question and one color.”

The passes themselves are not new. They are the same four passes from chapter five — structure, evidence, voice, gates — on a weekly cadence, at proposal scale. And a week per pass does not mean a week of passing. It means the daily ten minutes wears one color for seven days. Marcus never blocks an afternoon for this. He blocks ten minutes every morning, before the first meeting, and lets the week do the accumulating. Seven sittings add up to more sustained attention than any draft ever gets in one go, and unlike the blocked afternoon, the ten minutes actually happens.

I’ve tried to combine the passes, but I always miss something. The week I fix the voice, I forget to update the start date. The week I verify the dates, I leave in a sentence I’d never say out loud. So I now use Marcus’s calendar.

Week one is chapter five’s structure pass. What the proposal adds is the client’s order of questions. At Perimeter, the staffing section used to come after pricing, until Marcus moved it, because a client wants to know who will do the work before she sees what they cost. Last month he boxed a three-page company history that answered no question in the prompt. Each sitting, he reads standing up until the timer goes, draws a box around anything that answers a question nobody asked, and stops. The boxes wait for their own week.

Week two is the evidence pass, which Marcus calls proof once contracts are involved. Two monitors, sources on the left, proposal on the right, and each morning's ten minutes picks up at the line where yesterday's stopped. Top to bottom, no skipping around. This is the pass where you find the expired insurance certificate and the page reference that sends the reader to appendix C when the table is in appendix D. He once found a catering line item in a facilities bid, carried over from an older template, and he caught it in this week because he was not distracted by wording.

Week three is the voice pass, read aloud in an empty conference room, one stretch of pages per sitting. He once found a paragraph that began "Perimeter Logistics is pleased to submit this proposal in response to your stated need for." He cut everything after "Here is our answer." The cuts go down in black, and he doesn't second-guess them.

Week four is the gate pass in its business clothes: mechanics. Fonts, margins, signatures, page counts, file names. Rename the final PDF, and check that the attachments open on a computer that is not yours. Marcus once submitted a proposal with every page numbered one, but the client opened it anyway. He still checks pagination last, and he also checks that his own name is spelled right in the signature block. He's seen people catch everything in the body and miss the cover letter.

The calendar helps because pressure lies. With four weeks left, you tell yourself there's time to blend the passes. You'll do structure and proof together, then coast through voice, but with four days left, you do what's left. The separation isn't bureaucracy, and neither is the small clock. Ten minutes never gets cancelled, where an afternoon of review always does. So Marcus blocks the sittings on his calendar and treats them like client meetings he can't move. Four weeks of a ten-minute habit is what carried a proposal that outweighed his laptop; at no point did the proposal get a bigger clock, only more mornings.

Find a document you wrote last month, and give it one ten-minute sitting for structure. Mark the weak sections but don't rewrite them. Tomorrow, ten minutes verifying only the numbers. Read it aloud on the third day, because Friday is for commas, and see which pass catches the flaw that mattered most. A longer document gets more sittings, not longer ones.

Appendix: Cards, Templates and Where to Check

Appendix: Cards, Templates and Where to Check

Dana kept a single card above her monitor with tape that had browned at the corners. It was from the first time a vendor told her their platform reduced onboarding time by seventy percent. She'd believed the number until she couldn't find the footnote. Now the card held three checks. She ran them before every renewal call.

You can't trust your memory when someone is speaking quickly and showing a chart you aren't allowed to keep. That's why the tools in this appendix exist. They're not comprehensive. They're small enough to fit in a drawer or a single screen, because anything larger won't get used.

The card above Dana's monitor is the claim card. It holds three checks you run on any sentence that asks for money.

First, find the qualifier. Look for words like all, most, many, or some. The qualifier tells you whether the speaker's staking everything or hedging. Second, find the quantifier. This is the number, the percentage, or the multiple. If there isn't one, mark the space empty. Third, find the causal verb. These are words like reduced, caused, drove, and enabled. They pretend to prove a link between the product and the result.

If a pitch is missing any of the three, you don't have a claim. You have a story. Dana crosses the boxes with a pen. It takes ten seconds.

Marcus in procurement keeps a second card taped inside his notebook. He added it six months after Dana started using her claim card. The source card is for when the numbers sound right but the origin is missing.

Ask where the measurement came from. Was it the vendor's own customer success team, a third party, or the client's internal audit? Ask when. A two-year-old benchmark isn't evidence for next quarter. Ask who paid for the study. You don't need to reject sponsored research automatically, but you need to know the sponsor before you weigh it. Marcus says the card keeps him from nodding along to statistics that came from a firm's own marketing department.

Dana made the third card after a renewal went sideways. The vendor had proved their tool saved hours each week, but the case study involved a company ten times their size. The scope card checks whether the example matches your situation.

A result achieved at a twelve-hundred-person hospital may mean nothing for your eight-person clinic. A contract clause that's standard for enterprise software can break a small deployment. The scope card has two questions. What did they test this on? What did they leave out?

If the vendor can't answer both, you're buying on a hope.

Cards handle sentences. Templates handle documents.

When Marcus reviews a new vendor agreement, he doesn't start on page one. He drops the contract into a three-column sheet. The first column lists every promise the salesperson made in the deck. The second column lists the clause that backs it up, or he writes none found. The third column is for his own notes. If column two is empty at the end, the promise doesn't exist in the agreement. He sends the open items to the rep before he signs anything.

Dana uses a simpler template before she takes a call. She draws a line down a blank page. On the left she writes the vendor's claim in their exact words. On the right she writes what she'd need to see to believe it. She leaves the right side blank during the call. Afterward she emails her list to the rep. Requests phrased this way are harder to deflect than general skepticism.

Some facts you can verify while the caller's still on the line. Others you flag and handle later.

For public companies, filings are the ground floor. Revenue, headcount, and material contracts show up in the 10-K where press releases won't go. For product capabilities, the documentation beats the demo. Demos run on preloaded data with a smiling guide in the room. For legal terms, the state or federal registry tells you if the company exists. It also shows whether the firm has changed names twice in three years.

As a rule of thumb, if you can't find a source in five minutes of honest search, stop. Write it down as open and move on. A review that stalls on every footnote never reaches the clauses that can cost you money.

Pick one claim you accepted in the last month. Write the sentence down exactly as you heard it or read it. Run it through the three cards. Find the qualifier, the quantifier, and the causal verb. Name the source and the scope. If any check comes up empty, that's where your risk lives. Build a card for that gap and keep it where you'll see it before the next call.

Before you go

If *Ghost Words* changed how you handle something, a short review helps more than it looks like it should. It is how the next reader decides whether this is worth an hour of their time.

The most useful reviews name the one thing that worked. A sentence is plenty.

And if something did not work, if a check felt wrong or a chapter missed the point, that is worth more to me than praise. Say so.

About the author

Bandar Alsharfi has spent decades in enterprise technology, most of them helping large, regulated institutions adopt new systems and then live with the consequences. He is drawn to the parts of AI that do not demo well: governance, compliance, cost, vendor risk, and the question of who signs off when something goes wrong. During his career, Bandar has sat on both sides of that question, as the person recommending the system and as the person who has to answer for it.

He holds an Executive MBA and a degree in Information Systems, and is completing an MSc in AI Engineering. He writes at codedebate.com.

Also by Bandar Alsharfi

- **Who Told It to Do That?**: A Manager's Guide to AI Agents, for When the Answer Is You
- **It's Probably Fine**: The AI Incident Playbook for the Day It Very Much Is Not
- **Great Essay. Who Wrote It?**: Knowing Who Wrote What, When AI Drafts First and the Detector Is Guessing
- **We'll Firm That Up in Discovery**: Scoping, Pricing and Delivering AI Advisory That Reaches Phase 2
- **The Second Line Would Like a Word**: Shipping AI in Financial Services Before the Use Case Retires
- **The Chatbot Didn't Go to Your School**: Professional Judgement in the Age of the Confident Machine
- **Duly Noted**: The Five AI Decisions Your Board Actually Owns
- **Because the Robot Said So**: Asking the Follow-Up Question When the AI Sounds Sure
- **The Demo Was Flawless**: Buying Enterprise AI, from the Wrong Side of the Table
- **Are We Paying Too Much?**: Scaling AI FinOps in the Enterprise Jungle
- **Close Enough to Be Wrong**: Why AI Summaries and Meeting Notes Quietly Change What Was Actually Said — and the One-Minute Check That Catches It
- **Suddenly Everyone's an AI Expert**: A Small Business Owner's Guide to the Three Workflows That Actually Pay

www.ingramcontent.com/pod-product-compliance
Lightning Source LLC
LaVergne TN
LVHW010738061126
851857LV00014B/1541